Specification for Mapping at Scales between 1 : 1,000 and 1 : 10,000

The Royal Institution of Chartered Surveyors

Contents

Section		Page
	Introduction	i
1	Project Definition	1
2	Scope of Work	2
	2.1 Scale and Contour Interval	2
	2.2 Mapping Area	2
	2.3 Conventional or Digital Mapping	2
	2.4 Contractual Tasks	3
3	Materials to be Delivered — Bill of Quantities	4
4	Survey Methods and Quality Assurance	5
5	Aerial Photography (where applicable)	5
6	Projection, Grid and Height Datum	5
	6.1 Reference System	5
	6.2 Height Datum	5
7	Ground Control	6
	7.1 Control Diagram	6
	7.2 Plan Control	6
	7.3 Height Control	6
	7.4 Permanent Ground Markers	7
	7.5 Permanent Bench Marks	7
8	Map Detail	8
	8.1 Planimetric Accuracy	8
	8.2–8.9 Detail to be shown on Final Maps	8–10
9	Contours	11
	9.1 Accuracy	11
	9.2 Obstructions	11
	9.3 Steep Slopes	11
10	Spot Heights	12
	10.1 Accuracy	12
	10.2 Locations	12
	10.3 Additional Higher Precision Spot Heights	12
	10.4 Digital Terrain Model	12
11	Additional Information to be Supplied (including Field Completion of Photogrammetric Plots)	13
	11.1 Access and Damage	13
	11.2 List of Additional Information to be Supplied	13
	11.3 Special Features to be Supplied	13
12	Preliminary Plots	14
13	Presentation of Final Maps	15
	13.1 Sheet Size and Layout	15
	13.2 Grid	15
	13.3 Border	15
	13.4 Presentation	16
	13.5 Contours and Spot Heights	16
14	Digital Data	17
	14.1 Digital Data for Automated Plotting	17
	14.2 Digital Data for Entry to a Geographic or Land Information System	17
	14.3 Feature Codes	17
	14.4 Test Data	17
	14.5 Data Transfer Format	18
15	Special Requirements	19
16	Materials and Assistance to be Provided by the Client	19
17	Delivery Schedule	19
	Annexe A	20

Introduction

The purpose of the second edition of this specification remains the same as the first: to provide a general technical specification for contract mapping worldwide which can be modified as required by commissioning agencies and surveyors to meet the particular needs of individual mapping projects.

Experience and criticisms of the first edition, resulting from its widespread use around the world for seven years, have been used to clarify and update this second edition, which has been extended to include digital mapping.

Notes are provided on the left hand pages for guidance in selecting the appropriate options, and altering or omitting certain clauses in order to prepare contract specifications for particular mapping projects. These notes do not form part of the specification.

As far as possible, the products to be delivered are specified, not the survey methods to be used. This should assist commissioning agencies in selecting the scale, contour interval, accuracies, map content and graphical or digital products which are necessary for the particular project, conscious that costs escalate rapidly with increasing scale, closer contour interval, higher accuracies and more detailed map content. It also permits surveyors to select the most timely and cost-effective combination of ground or aerial, conventional or innovative surveying techniques to meet the requirements both rapidly and economically.

The final products may be supplied as hardcopy map transparencies, digital data for automated drafting, digital terrain models, or digital datasets for entry to geographic or land information systems.

This specification is intended to be used in conjunction with the RICS *Terms and Conditions for Survey Contracts* and the RICS *Specification for Vertical Air Photography* (second edition), both to be published in 1988. The RICS *Specification for Surveys of Land, Buildings and Utility Services at Scales of 1:500 and Larger* is also published by Surveyors Publications.

Comments on this revised specification continue to be welcomed for consideration in the preparation of future editions and should be addressed to:
The Land Surveyors Division
The Royal Institution of Chartered Surveyors
12 Great George Street, Parliament Square,
LONDON SW1P 3AD

Notes for Users

(These notes are provided for guidance and should be omitted from the specification.)

This specification is designed to be used as a contract document by entering appropriate details in the spaces provided, deleting Sections and numbered paragraphs which are not required, and adding Appendices or additional Sections and numbered paragraphs to include special requirements not covered in the document itself.

For the convenience of users with IBM-compatible wordprocessors, this document can also be purchased from The Royal Institution of Chartered Surveyors as a 5¼ inch diskette file in WordStar format and as a simple ASCII file on the same diskette for conversion to other wordprocessor formats. This will greatly simplify the task of modifying the standard document.

Where users make small but significant changes to numbered paragraphs in the standard document, it is suggested that these are listed at the beginning of the modified specification or indicated beside the text to ensure that the changes are not overlooked.

1 Insert details for project administration.

Section 1 Project Definition

NAME OF PROJECT: ....

PROJECT REFERENCE: ....

DATE OF ISSUE: ....

PURPOSE OF PROJECT: ....

COMMISSIONING AGENCY: ....

Address: ....

....

Telephone number: ....

Telex number: ....

Fax number: ....

Contract Officer: ....

Telephone extension: ....

SURVEYOR: ....

Address: ....

....

Telephone number: ....

Telex number: ....

Fax number: ....

Contract Officer: ....

Telephone extension: ....

2.1 Insert selected mapping scale.
Insert contour interval here and in Section 9 or delete line and whole of Section 9 if contours are not required.
Delete spot heights here and in Section 10 if not required.

2.2 Attach a map, or a diagram with boundary coordinates, to define the mapping area, and insert the approximate area.

2.3 For *conventional* mapping select one option from (a), (b) or (c) and delete all the others. Select the appropriate numbered paragraphs in Section 13 and delete those not required. Delete Section 14.

For *digital* mapping several of the options (c), (d), (e) and (f) may be required in combination, such as (c), (e) and (f). Select the appropriate numbered paragraphs from Section 14 *Digital Data* and numbered paragraph 10.4 *Digital Terrain Model* and delete those not required.

Section 2 Scope of Work

2.1 Scale and Contour Interval

Mapping is required at the scale of .... :
with *Contours* at a vertical interval of .... metres as specified in Section 9,
and with *Spot Heights* as specified in Section 10.

2.2 Mapping Area

The area to be mapped measures approximately .... square kilometres/hectares and is defined on the drawing or list of coordinates attached as Appendix......... .

2.3 Conventional or Digital Mapping

The mapping shall be supplied as:

Either (a) direct reproductions of photogrammetric instrument plots or ground survey plots without final drawing as specified in 13(a);

or (b) improved preliminary plots as specified in 13(b);

or (c) final maps of a high cartographic standard, as specified in 13(c), which may be prepared by manual plotting and drafting or by automated cartography. No digital data is required to be delivered;

and/or (d) digital mapping to produce digital data in a format suitable for automated plotting as specified in 14.1 and 14.5, together with final map sheets plotted from the digital data, as specified in 13(d);

and/or (e) digital mapping to produce digital data structured to the entry requirements of the geographic or land information system specified in 14.2 and 14.5, together with final maps derived from the digital data as specified in 13(d);

and/or (f) digital terrain model as specified in 10.4 and 14.5.

2.4 Delete items not required to be supplied by the surveyor and amend section numbers if changed.

2.4 Contractual Tasks

(Delete items not required to be performed by the surveyor)

The work involves the following tasks:

Tasks	Specified in Section
• aerial photography	5
• ground control	6 & 7
• permanent ground markers	7.4
• permanent bench marks	7.5
• detail mapping	8
• contours	9
• spot heights	10
• higher precision spot heights	10.3
• digital terrain models	10.4, 14.3 & 14.5
• additional information	
(including field completion of photogrammetric plots)	11
• special features to be supplied	11.3
• preliminary plots	12
• final maps reproduced from:	
preliminary plots	13(a)
improved preliminary plots	13(b)
drawings of high standard	13(c)
plots from digital data	13(d)
• digital data	
digital data for automated plotting	14.1, 14.3 & 14.5
digital data for entry to GIS/LIS	14.2, 14.3 & 14.5
digital terrain model	10.4, 14.3 & 14.5
• special requirements	15

3   Delete items not required and insert any additional deliverable products in 3.14 and Section 15.

4

3.15   Insert retention period and instructions for disposal. The surveyor may reasonably expect to charge a fee for storage beyond two years.

Section 3 Materials to be Delivered — Bill of Quantities

(Delete items not required)

Item	Quantity	Unit	Description
3.1	1	set	Original aerial film negatives.
3.2	....	set(s)	Contact prints of the aerial photography.
3.3	....	set(s)	Transparencies of index plot of the aerial photography.
3.4	....	set(s)	Station descriptions and lists of coordinates and heights of permanent ground markers constructed and existing permanent survey stations used in the survey.
3.5	....	set(s)	Station descriptions and list of heights of permanent bench marks constructed and existing bench marks used in the survey.
3.6	....	set(s)	True-to-scale transparencies of the preliminary plots reproduced in reverse on stable base material as specified in Section 12.
3.7	....	set(s)	Paper diazo copies of the preliminary plots.
3.8	....	set(s)	Paper proofs of the following: • master border and title block • diagram of proposed sheet layout and sheet numbering system • each final map sheet
3.9	....	set(s)	True-to-scale transparencies of the final maps reproduced as a positive, printed in reverse (wrong reading), on stable base material with matt drawing surfaces on both sides.
3.10	....	set(s)	Paper diazo copies of the final maps.
3.11	....	set(s)	Digital data as specified in Section 14.
3.12	....	set(s)	Check plot transparencies produced direct from the digital data.
3.13	1	set	Test data as specified in 14.4.
3.14			Special requirements as specified in Section 15.
3.15			**Working Materials** Original working materials, such as ground control results, plots and drawings, shall be retained by the surveyor for at least .... years after completion of the contract and may be disposed of thereafter, except where special instructions for disposal are given below: ....

4   Clients may wish to adopt the international quality assurance standards as applied to surveying and mapping ISO9001 (BS5750) when published.

5   Select option required or delete this Section if the work is to be done by ground survey methods.

6.1   Insert details.

6.2   Insert height datum.

Section 4   Survey Methods and Quality Assurance

The surveyor shall employ staff on this project who are experienced in the various tasks to be performed, and shall use techniques, equipment and materials which are capable of achieving the accuracies and standards specified for the final products.

The client shall be entitled to inspect the work in progress at any time and may request written details of the techniques, equipment and staff to be employed on the project.

Section 5   Aerial Photography
(Where applicable)

Either (a) the mapping shall be prepared from existing photography and either the original films, or prints and diapositives, shall be supplied to the surveyor;

or (b) the surveyor shall fly aerial photography of a scale and quality suitable for the preparation of the maps and other products specified under this contract, in accordance with the RICS *Specification for Vertical Air Photography* as modified and attached as Appendix ....

Section 6   Projection, Grid & Height Datum

6.1   Reference System
Ground control and mapping shall be related to:
Either (a) the National Grid/Universal Transverse Mercator Grid (UTM) in metres;

or (b) the grid defined below in metres:

- projection ....
- ellipsoid ....
- datum ....
- coordinates of origin ....
- false coordinates of origin ....
- central meridian/standard parallel ....
- scale factor at ....

6.2   Height Datum
All control points, heights and contours shall be related to the .... height datum in metres.

7   Delete options not required, and modify remainder of Section 7 as necessary.

7.2   Standard accuracies of control for mapping are given, which can be amended if required to conform to local practice or special engineering requirements.

Appropriate methods of adjustment of old and new control data should be agreed with the surveyor.

7.3   To be read in conjunction with 7.5

Appropriate methods of adjustment of old and new control data should be agreed with the surveyor.

Section 7   Ground Control

Either (a) existing control available to the surveyor shall be used;

or (b) the client shall establish ground control to a pattern and accuracy to be agreed with the surveyor;

or (c) the surveyor shall establish ground control as specified in 7.1 to 7.5. The density of control points shall be sufficient to achieve the mapping accuracies specified in 8.1, 9.1, 10.1, and (where applicable) 10.4.

7.1 Control Diagram

The client may request a diagram to be submitted for approval showing the planned density and pattern of plan and height control points.

7.2 Plan Control

Plan control for photogrammetric mapping may be established by signalisation (targeting or pre-marking) before photography or by photo-identification of existing features after photography, or by a combination of both.

New plan control shall be established throughout the mapping area to an accuracy of better than one part in 20,000 as determined by loop closures or other redundant observations.

The coordinate values of adjacent control points shall be in sympathy with each other to a root mean square error of better than ±0.06 mm at map scale. (90% of a representative sample of points shall be in sympathy with adjacent points to better than ±0.1 mm at map scale.)

Where mapping is to be based on existing control of lower accuracy, any new points shall be established and adjusted in sympathy with the existing net to comparable accuracy.

7.3 Height Control

Where new height control points are established, adjacent points shall be in sympathy to better than one tenth of the specified contour interval, and all height control points shall be in sympathy with existing bench marks or a reference bench mark to better than one third of the specified contour interval.

Where mapping is to be based on existing bench marks of lower accuracy, height control points shall be established and adjusted in sympathy with the existing bench marks to comparable accuracy.

7.4 Delete options not required.

Alternative designs may be inserted instead.

Attach a map showing the approximate locations of PGMs as an Appendix, or insert a description of the general density and types of locations required.

7.5 Delete options not required.
Alternative designs may be inserted instead.
Attach a map showing the approximate locations of PBMs as an Appendix, or insert a description of the general density and types of locations required. Standard accuracies of height control for mapping are given, which can be amended if required to conform to local practice or for special engineering requirements.

7.4 Permanent Ground Markers

Either (a) ground control stations are not required to be permanently marked on the ground;

or (b) main survey stations such as traverse stations and bench marks shall be permanently marked on the ground by monuments constructed to the designs shown in Annexe A. Photopoints are not required to be permanently marked;

or (c) permanent ground markers (PGMs) shall be constructed to the designs shown in Annexe A in the positions or pattern given in Appendix ....

All permanent ground markers constructed under this contract, and existing permanent survey stations used for the mapping, shall be plotted and numbered on the final maps, and station descriptions with distances to reference objects and a list of coordinates and heights shall be supplied to the client.

The construction of permanent ground markers may be modified to suit local conditions with the agreement of the client. In unstable ground such as swamp or sand dunes it may not be feasible to construct permanent ground markers.

7.5 Permanent Bench Marks

Either (a) permanent bench marks are not required;

or (b) permanent bench marks shall be constructed to the designs shown in Annexe A in the positions or pattern given in Appendix ....

The height difference between any two permanent bench marks shall not be in error by more than $\pm(15\sqrt{k})$ mm where k is the distance between them in kilometres.

8.2 Where ground survey methods are used, ground-lines of buildings will normally be surveyed. Where air survey methods are used, roof-lines will normally be plotted from the aerial photography but can be replaced by ground-lines by field completion on the ground if specified in 11.2. The difference between roof-lines and ground-lines is rarely significant at scales smaller than 1:1,250 except where buildings have roofs, balconies, or upper floors overhanging by more than 0.6 metres. See 11.2.

Section 8 Map Detail

Topographical features shall be surveyed and plotted on the final maps as defined in 8.1 to 8.9, together with the additional information listed in Section 11.

Where ground survey methods are used, the features required in Sections 8 and 11 may be collected simultaneously and proofs of the final maps submitted for approval.

Where air survey methods are used, the features described in Section 8 which are clearly apparent on the air photographs shall be plotted and only the features listed in Section 11 are required to be supplied by field completion on the ground.

8.1 Planimetric Accuracy

Grid lines and control points shall be drawn to an accuracy better than ±0.3 mm maximum tolerance.

An additional tolerance for shrinkage of stable base materials shall be permitted. Provided final transparencies are stored carefully at a temperature of around 20°C and a relative humidity of around 50%, stable base materials should remain dimensionally correct within ±0.3 mm in one metre.

Well-defined points of detail shall be plotted in their true positions at map scale to better than ±0.3 mm root mean square error, when coordinates are scaled off the map from the nearest grid lines and compared with coordinates determined by precise measurement on the ground from the nearest control point. (90% of a representative sample of well-defined points shall be within ±0.5 mm.)

Detail to be Shown on the Final Maps

8.2 Buildings and Structures

Permanent buildings larger than 6 square mm at map scale shall be shown by roof-lines or ground-lines.

Smaller buildings may be generalised or omitted as appropriate. Ruins, partially demolished buildings, buildings under construction and other structures shall be shown in outline. Glasshouses larger than 8 square mm at map scale shall be distinguished by cross-hatching.

8.3 Boundaries

Walls, hedges, fences and similar field boundaries shall be shown by single lines representing the centre of the physical boundary, except for walls more than 1 mm in width at map scale which shall be shown by double lines, and hedges more than 3 mm in width at map scale which shall be shown by conventionalised canopy symbols.

Administrative boundaries shall not be shown unless specified as a special requirement in 11.3.

8.6 Additional features required may be inserted in 11.2 or 11.3.

8.7 Delete options not required.

Insert a contour or height value related to the established height datum, such as mean sea level or chart datum.

8.4 Roads, Tracks and Footpaths

Road edges or kerbs shall be surveyed, except at scales smaller than 1:5,000 where widths may be generalised according to the road category.

Edges of tracks and footpaths shall be surveyed except where indistinctly defined or less than 2 mm in width at map scale, where tracks shall be shown by double lines of standard width and footpaths by single lines.

Drives and tracks in private properties shall be shown only where more than 50 metres in length.

8.5 Railways

Railway tracks shall be shown either by the gauge width or by conventional symbols.

Railway stations, buildings, bridges and level crossings shall be shown, but other railway installations shall be omitted unless specified as additional features in Section 11.

8.6 Transmission Lines, Pipelines, Masts and Poles

Pylons and masts shall be shown by conventional symbols except where larger than 4mm across at map scale, where they shall be surveyed to scale.

Surface pipelines shall be shown by conventional symbols.

Electricity and telephone poles outside urban areas shall be shown at 1:2,500 and larger scales.

8.7 Water, Drainage and Coastal Features

Rivers, streams, canals and ditches more than 2 mm in width at map scale shall be shown by double lines, and narrower features by single lines.

Intermittent streams and wadi beds where significant shall be shown by broken lines.

Rivers and other water features obscured by trees or undergrowth shall be shown by broken lines to indicate their approximate alignment.

The water level of rivers, lakes, ponds, lagoons and reservoirs shall be shown by the water line at the time of photography or of ground survey.

Wells, springs, waterfalls, dams, weirs, sluices, locks and fords shall be surveyed in outline or indicated by symbols or annotations where the features are too small to plot at the scale of mapping.

The sea shoreline shall be shown by:

Either (a) the water line at the time of photography or ground survey;

or (b) the approximate high water mark;

or (c) the .... level in metres related to .... height datum.

Major features such as mud, sand, shingle, boulders, rocks, cliffs, swamps and marshes shall be indicated appropriately by symbols or annotations.

Docks, piers, jetties, slipways, harbour walls, fixed cranes, breakwaters, groynes, and lighthouses shall be surveyed in outline or shown by symbols as appropriate to the scale.

8.8 Insert any additional features or classifications required in 11.2 or 11.3.

8.9 Delete options not required.
Insert detailed name collection requirements in 11.3.

8.8 Terrain, Vegetation and Land Use Classifications

The representation of major types of terrain, vegetation and land use shall be limited to simple classifications of significant and extensive topographical features. These features shall be shown by symbols or annotations.

Terrain features to be shown shall include rock outcrops, cliffs, sand dunes, swamps and marshes.

Vegetation and land use features to be mapped shall be limited to major categories of woodland, bush, scrub, cultivation, orchards and plantations, which shall be outlined and identified by conventionally spaced symbols or annotations. Individual tree trunks are not required to be shown (unless specified in 11.2).

At scales larger than 1:2,000 woodland and large isolated trees shall be shown by the extent of the canopy.

At 1:2,000 and smaller scales woods shall be outlined and indicated by conventionally spaced symbols. Scattered trees shall be represented by scattered symbols, and only prominent isolated trees shall be shown individually.

Man-made features to be shown shall include open-cast mines, quarries, tips, cemeteries, parks and recreation grounds.

Additional classifications shall be mapped only as listed in 11.3.

8.9 Names and Annotations

Names of places, districts, streets and prominent public buildings shall be shown in English or ...., or both on the final maps:

Either (a) taken from existing maps;

or (b) supplied by the client;

or (c) collected on the ground as specified in 11.3.

9   Delete this Section if contours are not required.
        Ensure contour interval corresponds with that stated in 2.1.

9.2   Delete option not required.

Section 9   Contours

(Delete if not required)

Contours shall be shown at a vertical interval of .... metres.

9.1 Accuracy

Contours shall be correct to a root mean square error of better than one third of the contour interval where a representative sample of points on contour lines is checked by precise measurement from the nearest control point. (90% of a representative sample shall be correct to better than half the specified contour interval.)

Any contour which can be brought within this vertical tolerance by moving its plotted position in any direction by not more than 0.5 mm or one tenth of the horizontal distance between contours, whichever is the greater at map scale, shall be considered acceptable.

9.2 Obstructions

Where, because of trees, vegetation or other obstructions, the ground is not visible on the air photographs or line of sight for ground survey is restricted:

Either (a) contours shall be shown as broken lines to indicate that the accuracy stated in 9.1 cannot be guaranteed;

or (b) the contour accuracy stated in 9.1 shall be maintained by height measurement on the ground, provided that the client obtains authority and accepts liability for damage to crops and vegetation caused during clearing or survey.

9.3 Steep Slopes

On steep slopes, intermediate contours may be omitted where they are generally closer than 1.5 mm apart at map scale.

10 Delete this Section if spot heights and a digital terrain model are not required.

10.2 Delete if not required, or insert locations and densities of additional spot heights.

10.3 Delete this numbered paragraph if higher precision spot heights are not required, otherwise insert a map showing the areas where higher precision spot heights are to be supplied with a description of the locations and density.

The accuracy specified may be varied to suit the requirements of the project.

10.4 Delete this numbered paragraph if a digital terrain model is not required, otherwise insert a map showing the areas where a digital terrain model is to be supplied with a specification of the accuracies, pattern and density of points. The digital data format should be specified in 14.5.

Note: The UK Department of Transport *Specification for Topographical Surveys*, January 1984, includes specifications for both regular grid, triangular and string digital terrain models for use with MOSS and BIPS3 highway design systems.

Section 10   Spot Heights

(Delete if not required)

10.1 Accuracy

Spot heights shall be correct to a root mean square error of better than one quarter of the specified contour interval where a representative sample is checked by precise measurement from the nearest control point. (90% of a representative sample shall be correct to better than four tenths of the specified contour interval.)

10.2 Locations

Spot heights shall be shown in the following positions, except where the ground is obscured by vegetation or other obstructions:
- at salient points such as hilltops, bottoms of depressions and saddles
- at significant changes of gradient along the centre line of through roads, generally at intervals of between 50 and 100 mm at map scale
- in flat areas (where the horizontal distance between contours generally exceeds 50 mm at map scale) at intervals between 50 and 100 mm at map scale
- at the locations or densities specified below:

....

....

10.3 Additional Higher Precision Spot Heights
(Delete if not required)

Higher precision spot heights shall be supplied on the final maps in place of standard spot heights in the areas and locations defined in Appendix ....

Higher precision spot heights shall be correct to a root mean square error of better than ±0.01 metres where a representative sample is checked by precise measurement from the nearest bench mark. (90% of a representative sample shall be correct to better than ±0.017 metres.)

10.4 Digital Terrain Model
(Delete if not required)

A digital terrain model shall be supplied of the areas indicated in Appendix .... and to the patterns, density, and accuracy defined in Appendix ....

(The data transfer format for entry to the modelling system is specified in 14.5.)

11 Delete options not required, and delete this Section if no additional information is required.

When old aerial photography is to be used, this numbered paragraph may be modified to specify that the mapping is to be updated by field completion either of all map features or of a selected list of features.

11.2 Delete items not required and expand list as necessary, taking note of the scale of mapping being specified.

Ground-lines or roof-lines. Delete this sub-paragraph if roof-lines plotted by air survey are acceptable.
See note 8.2.
Where air survey methods are used, roof-lines will normally be plotted from the aerial photography but can be replaced by ground-lines by field completion on the ground if specified in this numbered paragraph. The difference between roof-lines and ground-lines is rarely significant at scales smaller than 1:1,250 except where buildings have roofs, balconies, or upper floors overhanging by more than 0.6 metres.

11.3 Insert any additional features or information, not normally shown on topographical maps, which are required to be surveyed and plotted on the final maps.
Delete numbered paragraph if not required.

Section 11 Additional Information to be Supplied

(Including field completion of photogrammetric plots)

Either (a) no additional information is required;

or (b) the client shall plot the additional information required on copies of the preliminary plots and return these to the surveyor to incorporate into the final maps;

or (c) the surveyor shall collect the additional information described in 11.1 to 11.5 and incorporate it into the final maps. Where air survey methods are being used, changes which have occurred since the date of photography are not required to be shown.

11.1 Access and Damage

The work shall be limited to collecting information which can be acquired without entering private property or restricted areas, except where permissions to enter such areas are obtained by the client.

Where damage may be caused to trees, crops, or other obstructions in order to gain access or to clear survey lines, prior agreement must be reached with the client, who shall obtain authorisation and accept liability for any necessary damage caused.

11.2 List of Additional Information to be Supplied

- Names of districts, towns, villages, rivers, lakes, physical features, major roads and public buildings
- Classification of roads, tracks and surface pipelines
- Simple classification of major types of terrain such as rock, salt flats (sabkah), sand dunes and swamp
- Simple classification of major types of vegetation and land use such as woodland, scrub, grassland, cultivation and plantations
- Simple annotation of man-made features such as mines, tips, quarries, parks, cemeteries and car parks
- Bridges, culverts, distance markers, buried pipeline markers, gasometers, wells and other major landmarks
- Letter boxes, telephone call boxes, electricity sub-stations, pylons of power lines over 11,000 volts and fire hydrants *only at 1:2,500 and larger scales*
- Archways in buildings where more than three metres in width and visible from public access *only at 1:2,500 and larger scales*
- Poles of powerlines over 11,000 volts and gates more than three metres in width *only at 1:1,250 and larger scales*
- Ground-lines (plinths) of buildings shall be shown on the final maps instead of roof-lines *only at 1:1,250 and larger scales*

11.3 Special Features to be Supplied

The following special features shall be surveyed and shown on the final maps:

....

....

12 Insert period required by client to check preliminary proofs and return to the surveyor with corrections and allow the same period in the delivery schedule in Section 17.

Section 12   Preliminary Plots

Where specified in 3.6 and 3.7, transparencies and paper copies of the preliminary plots shall be delivered as proofs before field completion and preparation of the final maps.

One set of paper proofs shall be returned to the surveyor within .... weeks of receipt showing any corrections which the client requires to be incorporated in the final maps.

13   Delete options not required.

13.1   Delete option not required.
       Insert sheet size and layout or leave the design to the surveyor.

13.2   Delete option not required.

       Graticule cuts may be inserted if required.

13.3   Delete optional items not required.

Section 13   Presentation of Final Maps

Either (a) the final maps shall be reproduced direct from the preliminary plots as specified in Section 12 and no additional cartographic drawings are required;

or (b) the final maps shall be reproduced from the preliminary plots which shall be completed to a neat standard with typeset or stencilled names, annotations, grid and height values;

or (c) the final maps shall be reproduced from drawings of a high cartographic standard, prepared to the specifications given in 13.1 to 13.5, by drawing in ink or scribing or automated plotting from digital data or a combination of methods;

or (d) the final maps shall be derived from the digital data by automated plotting and produced to the standards specified in 13.1 to 13.5.

13.1 Sheet Size and Layout

Either (a) the size and layout of the final map sheets is given in Appendix .... ;

or (b) the final map sheets shall be of generally uniform dimensions, not exceeding A0 size (841 mm × 1189 mm) overall. Oversized sheets, outriggers, and insets may be used only with the approval of the client. Large blocks of sheets shall be aligned parallel to the grid and butt joined, with the sheet edges coinciding with round figure grid values.

For irregularly shaped areas and route surveys, the sheets may be arranged skew to the grid; and where there is a change of orientation between adjoining sheets the joins shall be indicated by cut lines without duplicating detail in the overlaps.

13.2 Grid

The grid shall be drawn across the face of the maps either as continuous lines or intersections at 100 mm intervals for scales such as 1:1,000 and 1:5,000, or at 80 mm intervals for 1:1,250 and 1:2,500 scales.

13.3 Border

The following information shall be shown in the margin of each sheet:
- sheet number
- scale as a representative fraction
- grid values
- contour interval and height datum
- compilation note
- index to adjoining sheets
- client's name
- surveyor's acknowledgement

The following optional items may be added if requested:
- legend
- north point
- scale bar
- title block
- projection, ellipsoid, datum, grid
- copyright statement

13.4 Delete option not required.

Commissioning agencies who have adopted a standard style of map presentation should include their own drawing specification and sample map. Others should require the surveyor to propose a suitable style of presentation and submit a legend and sample sheet of a similar project at the same scale for approval.

13.5 Delete this numbered paragraph if contours and spot heights are not required.

Insert index contour interval, usually every fourth or fifth:
2 m or 2.5 m indices for 0.5 m contours
    5 m   1 m
    10 m   2 m
20 m or 25 m   5 m
    50 m   10 m

13.4 Presentation

Either (a) the style of presentation of the final maps shall conform to the drawing specification or sample map attached as Appendix .... ;

or (b) the surveyor shall select conventional signs, line styles and widths, type founts and sizes suitable for the particular project, and shall submit a legend and sample map sheet of this or a similar project for approval by the client before mapping commences.

Names, annotations and values shall be typeset, stencilled or computer generated.

13.5 Contours and Spot Heights

Contours, where specified in Section 9, shall be produced to a high cartographic standard, with contour values reading up the slope at a density sufficient to identify all contours without ambiguity. Thicker index contours shall be shown at multiples of .... metres. Steep slopes shall be shown by contours or by slope or cliff symbols as appropriate.

Depression contours shall be distinguished either by an arrow pointing downhill, or by ticks on the lower side of the bottom contour, or by a spot height value at the lowest point.

Spot heights, where specified in Section 10, shall be shown on the detail separations. Higher precision spot heights, where specified in 10.3, shall be distinguished by additional decimal places, or by bolder figures.

14 Delete this Section if digital data is not required. Otherwise delete options not required. Option (c) could be selected alone or in combination with (a) or (b).
Retain the relevant numbered paragraphs 14.1 or 14.2 and/or 10.4 plus 14.3, 14.4 & 14.5 as indicated.

14.1 Delete if option 14 (a) is not required.

14.2 Delete if option 14 (b) is not required.
For this option it is essential to supply full details of the client's information system and the structure and format of the data to be supplied.

14.3 Delete option not required.
Commissioning agencies who have adopted a standard feature coding system should insert this in the Appendix. Others should require the surveyor to submit a list for approval.

14.4 Insert period required by client to read test data into system and approve or return any comments to the surveyor, and allow same period in the delivery schedule in Section 17.

Section 14   Digital Data

(Delete if not required)

Either (a) digital data suitable for automated plotting of the map sheets shall be supplied as specified in 14.1, 14.3, 14.4 & 14.5;

or (b) digital data formatted and structured for entry to the named geographic or land information system shall be supplied as specified in 14.2, 14.3, 14.4 & 14.5;

and/or (c) a digital terrain model shall be supplied as specified in 10.4, 14.3, 14.4 & 14.5.

14.1 Digital Data for Automated Plotting

The digital map data used to plot the final map sheets shall be supplied in the format specified in 14.5 to enable the client to replot the map sheets using the same line styles and symbols on a suitable automated plotting system. Data files shall correspond to map sheets at the acquisition scale, and shall be edge matched between sheets and files.

14.2 Digital Data for Entry to a Geographic or Land Information System

The digital map data used to prepare the final maps shall be restructured and supplied as input to the geographic or land information system or graphic database system as defined in Appendix .... (Full details of the data structure and input format must be supplied in this Appendix.)

The data shall be supplied as a continuous database in the format specified in 14.5. with no mathematical gaps or overlaps between files or sheet edges or in continuous features.

14.3 Feature Codes

Either (a) feature codes and levels shall conform to the list attached as Appendix .... ;

or (b) the surveyor shall submit a list of feature codes and levels to the client for approval before mapping commences.

14.4 Test Data

Prior to the delivery of the final digital data, a small sample of the map data or digital terrain model or comparable data shall be submitted to the client in the specified or proposed format described in 14.5, which the client shall test and approve as the format for data delivery or return with comments within .... weeks.

14.5 Delete if no digital data is required.
Otherwise complete the appropriate details listed and add any other specifications necessary to define the data format required.

Note: The UK National Transfer Format (NTF) or Standard Digital Survey Format (SDSF) or other national and international transfer formats which provide more detailed specifications may be inserted as an Appendix.

Cartridges etc:

14.5 Data Transfer Format

The surveyor shall supply digital map data suitable for entry to the client's system described below:

Hardware
- computer ....
- plotter ....

Software
- operating system and version ....
- graphics package ....
- geographic information system ....
- digital terrain modelling system ....

The data shall be supplied in the following format and media:

either (a) Magnetic tapes

- sets of tapes ....
- maximum length ....
- number of tracks ....
- density (bpi) ....
- mode (NRZI/PE) ....
- code (ASCII/EBCDIC) ....
- record size (bytes) ....
- block size (bytes/block) ....
- parity (none/odd/even/mark) ....

or (b) Floppy disks

- diameter ....
- single or double sided ....
- density (single/double/quad) ....
- sectoring (soft/hard) ....
- formatting instructions ....
- code ....

or (c) Other (specify) ....

15   This Section should be used to specify any special requirements to be supplied by the surveyor in addition to the mapping. Delete Section if not required.

17   The following delivery times may be used as a guide for small projects of less than 10 map sheets and larger projects of around 100 map sheets.

Item	Small project Weeks (Cumulative)	Larger project Weeks (Cumulative)
Air photography	4	12
Ground control	8	18
Preliminary plots	12	24–30
Return of proofs by client	15	28–34
Final maps	20	32–38
Digital data	20	36–40

Section 15   Special Requirements

Special requirements and products to be supplied are specified in Appendix
....

(Specifications for items such as aerial triangulation results, photomosaics, orthophotomapping, profiles and derivative mapping at smaller scales may be inserted in this Section as required.)

Section 16   Materials and Assistance to be Provided by the Client

The client shall provide:
- assistance in obtaining flying permits, visas, work permits, access to land, and other necessary authorisations
- all instructions and information necessary to complete the work
- corrected proofs of maps, sheet layouts, borders, and test data with instructions to proceed with the subsequent stages of map production, within the agreed turn-round periods given in Section 12 and 14.4.

Section 17   Delivery Schedule

The work shall be delivered progressively from the date of signature of the contract or the authorised start date, whichever is later, within the following schedule:

Item	**Weeks (cumulative)**
• contact prints and index plots of aerial photography	....
• ground control results	....
• preliminary plots and test data	....
• return of proofs by client	....
• final maps	....
• digital data	....
• other products (specify)	....

The surveyor shall advise the client periodically on the progress of the work and notify him of any anticipated delays in the completion dates.

Summary of Appendices and Insertions

Check list of insertions to be completed and Appendices to be attached specification document or deleted if not required:

Numbered
paragraph

1	Project definition details
2.1	Scale and contour interval
2.2	Mapping area, map or diagram
3	Bill of Quantities
3.15	Instructions for disposal of working materials
5(b)	Specification for vertical air photography
6.1	Projection, datum and grid
6.2	Height datum
7.4(c)	Permanent ground marker design drawings and locations
7.5(b)	Permanent bench mark design drawings and locations
8.7(c)	Sea level
8.9	Names and annotations
9	Contour interval
10.2	Spot height locations
10.3	Higher precision spot heights location map
10.4	Digital terrain model location map and specification
11.2	Additional information to be supplied
11.3	Special features to be supplied
12	Period for return of preliminary plots
13.1(a)	Size and layout of final map sheets
13.4(a)	Drawing specification or sample map
13.5	Index contour interval
14.2	Details of GIS/LIS and structure and format of data
14.3(a)	Feature codes
14.4	Period for approval or comments of test data
14.5	Details of data transfer format
15	Special requirements
17	Delivery schedule

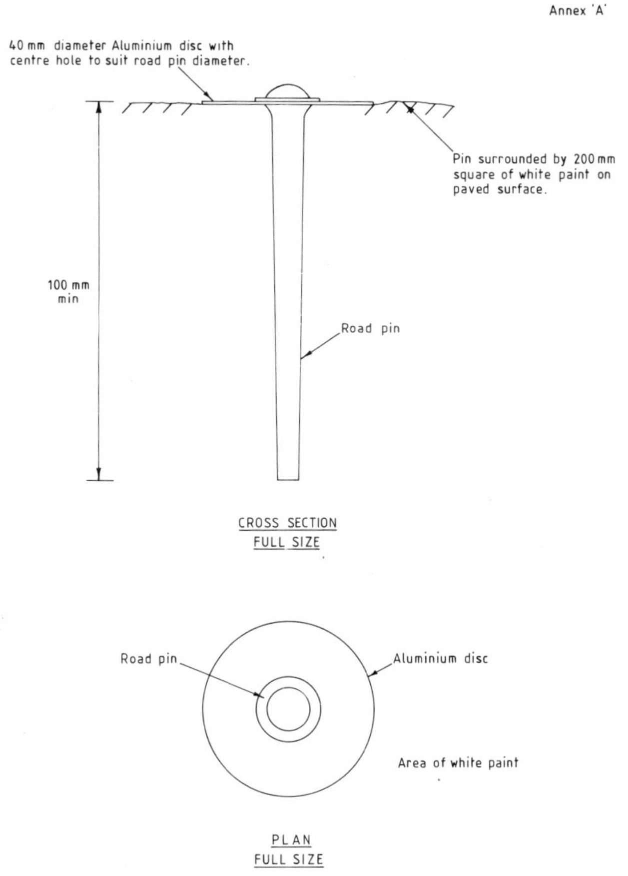

Annexe A

Permanent ground marker type 1 for dense very stable paved surfaces.

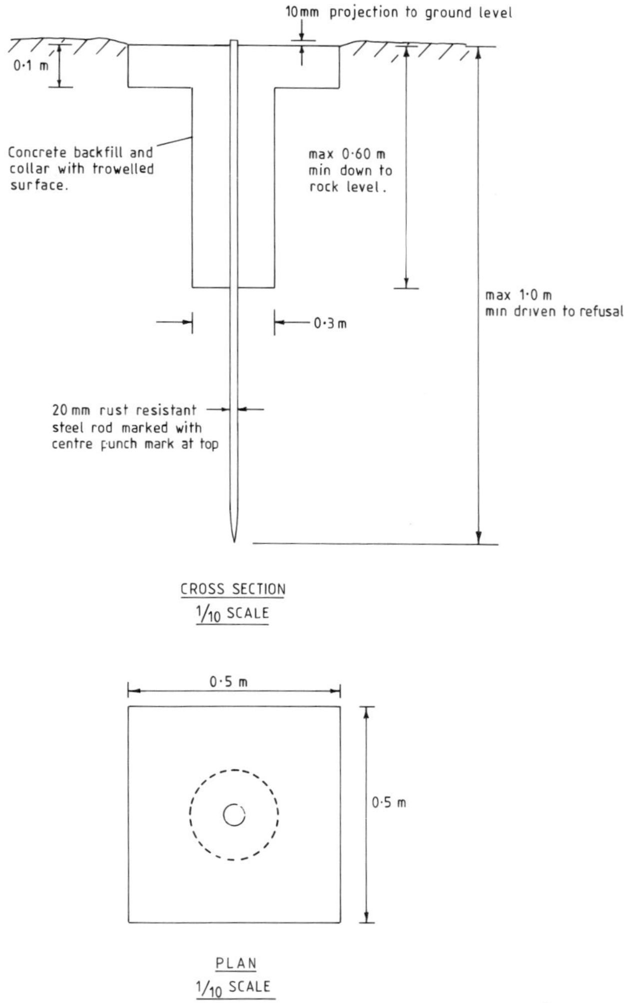

Permanent ground marker type 2 for non agricultural sites and unpaved surfaces.

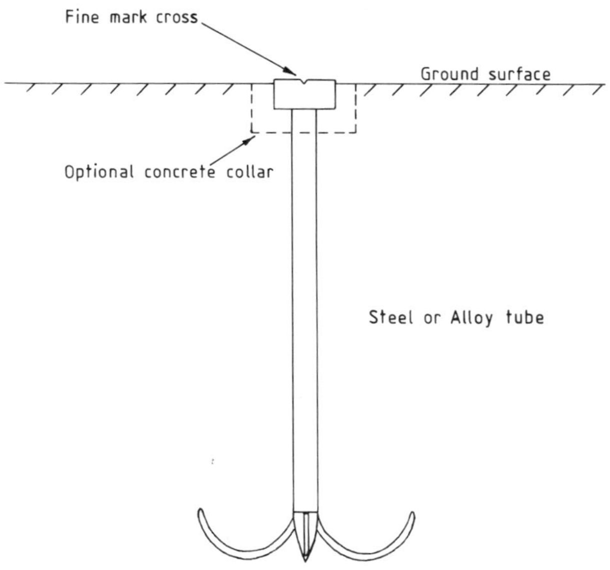

Permanent ground marker type 3 for soft surfaces.